RENE HEDLEY

CREATIVE RECYCLING SIDE

The Ultimate Guide on Recycling, Discover Effective Ways to Recycle So We Can Save the Earth and Create a Better Future For Our Children

Descrierea CIP a Bibliotecii Naţionale a României
RENE HEDLEY
   CREATIVE RECYCLING SIDE. The Ultimate Guide on Recycling, Discover Effective Ways to Recycle So We Can Save the Earth and Create a Better Future For Our Children / Rene Hedley – Bucharest: Editura My Ebook, 2021
   ISBN

RENE HEDLEY

CREATIVE RECYCLING SIDE

The Ultimate Guide on Recycling, Discover Effective Ways to Recycle So We Can Save the Earth and Create a Better Future For Our Children

My Ebook Publishing House
Bucharest, 2021

TABLE OF CONTENTS

Foreword 7

Chapter 1: **Recycling Basics** 9

Chapter 2: **Choose Wisely When Buying** 13

Chapter 3: **Donate Or Sell Items Instead of Throwing Away** 17

Chapter 4: **Reusable Bags** 20

Chapter 5: **Learn What Is ecyclable In Your Area** 23

Chapter 6: **Learn The Rules In Your Area** 26

Chapter 7: **Separate Your Trash** 29

Chapter 8: **Buy Recycled Items** 31

Chapter 9: **What To Avoid** 33

Chapter 10: **The Benefits Of Recycling** 35

FOREWORD

If you think that you are the only one who gets old, then, better think again. Even Earth itself has its own age and as the years go by, this age adds up making the world grow older and older, slowly but surely.

As one of the billions of people living in this world, one of your responsibilities is to ensure that even if the Earth grows older, it will not be affected by the drastic changes that can make its condition worsen. It is not a secret that the continuous piling up of waste has always been one of the serious issues that all countries need to deal with.

In this book, you will be further taken into the world of recycling for you to get to learn how this rather simple method can bring great changes to the whole world, specifically to your children and their future children.

CHAPTER 1

RECYCLING BASICS

Synopsis

What do you know about recycling? Are you aware of the things that can and cannot be recycled? What can be done to the recyclable things? Learn about the basics of recycling in this chapter.

A Glimpse Into the World of Recycling Surely, you know that everyone must recycle but majority of you are probably still a bit unaware of how and what to recycle.

Most people simply assume that they are doing the right thing by tossing their recyclable to the bin. Even though this is considered as a crucial step for this process, you need to be mindful of several additional things here. Here, you will explore how you can actually be a conscientious and effective recycler that will then allow you to keep the planet, including the oceans, clean and green.

The very first question that needs to be answered is what are the things that can be recycled? In general, the recyclables are being divided into 4 major categories: Plastics, Paper, Metals, and Glass. Depending on the recycling laws in your area, you may or may not need to divide the recyclables that you have accordingly.

What materials belonging to these categories should be recycled and how are they recycled?

Plastics

Majority of the plastics which include but are not limited to bags, containers, and bottles are recyclable. But this is only when these have been rinsed out, cleaned as well as emptied properly.

For the plastic bottles, first you need to remove the cap. Then, rinse and empty the bottle before flattening it and replacing the cap.

Keep in mind that when recycling the "6 pack rings," it is important to have them cut up for preventing any kind of tragedy in the marine life. You will definitely not want to be the one to blame for injuring or even killing innocent creatures.

Remember that toys, airbags, styrofoam and other kinds of plastic container that contain toxic material like motor oil cannot be recycled. The perfect way for dealing with such plastics is either to lessen if not limit the consumption of the products that are packaged in these or you can also search for more creative ways of reusing them.

Paper

Majority of the forms of paper items include cardboard, newspaper, plates and paper cups (rinsed out), magazines and books can be placed in recycle bins. Bear in mind that shredded paper will be separated from others. Also, remember that soiled paper, facial tissues together with the greasy pizza boxes are not recyclable. The best thing is that these things can be composted.

Metals

Steel, aluminum and tin are among those metals which are recyclable. Make sure, however, that all the cans will be emptied and cleaned. You cannot recycle hangers, aluminum foil, paint containers, scrap metals, and Helium containers. You can inquire in your place which will take such things off your hands for them to be placed in good use.

Glass

Jars, glass bottles and the containers that can be put in the recycling bin. However, you need to be aware that you need to put broken glass in the garbage bin or utilize this as a kind of art metal, not for recycling purposes.

Items including dishware, mirrors, glasses and cups, ornaments and decoration cannot be recycled. If you want to turn this kind of garbage to something useful in a fun, effective and creative way, you can use them as crafts and arts materials.

Through educating and informing your children about recycling's dos and don'ts, you can ensure that the future will still see a greener and cleaner planet.

CHAPTER 2

CHOOSE WISELY WHEN BUYING

Synopsis

One of the many factors that can affect the number of waste that are being piled up every single day is the way people purchase the things that they will use for their day to day activities. Discover how important it is to be wise when buying things and how it can help lessen the wastes that should be recycled.

The initial steps on the hierarchy of waste are to stay away from the need of collecting waste as well as to lessen the amount of waste being produced.

Through the reduction of the waste amount, you will also be able to reduce the number of waste that will need to be disposed of or recycled.

Be a Wise Buyer – Avoiding and Reducing Waste
Waste Avoidance

It is very simple to avoid waste and never take those things which you do not need or will not use. Plastic bags are among the unnecessary wastes that are very common today. Every time you shop, you will notice that most shops are providing plastic bags to carry home what they purchased home. However, if you will only buy a few things, it can be a wiser idea to avoid using plastic bags. In case you will buy a lot of things, it will be better to just take a reusable bag with you for carrying home the things that you purchased.

One more way for avoiding waste will be to avoid purchasing products that you do not need. This is where being a wise buyer comes in. A lot of people today purchase toys, electronics, and clothing even when the toys, electronics and clothing that they have at home can still be used. Before you buy anything new, it is crucial to consider if this item is needed or you only wanted to have it. One great way for getting new items is through trading with your friends.

Most people pack launches that they bring with them to school or work. It is a great way for avoiding waste. In place of putting your lunch in paper bag, you can use a reusable lunch box or take a lunch bag instead. For your snacks and sandwiches, you can use small and reusable containers. Try to avoid purchasing the prepackaged snacks such as biscuits that are individually wrapped or muesli bars since this kind of packaging will only generate more waste. A good alternative will be to purchase larger food packages and place them inside tiny, reusable containers.

Waste Reduction

It is also very simple to reduce waste. One great way for reducing waste is through reducing package. Most of the things that you purchase at stores are being covered in several packaging layers. Is this needed? As a wise buyer, you might as well consider purchasing those items that have lesser packaging. Majority of the items, like lentils, nuts and dry beans can be purchased in large quantities or in bulk. Instead of purchasing a box with a bean bag inside, for instance, you might as well purchase a bag that has large amounts of beans.

One more way of reducing waste, purchase products that you can reuse or refill, a good example of which is pens. You can find pens that are not cannot be opened or refilled. For this, it will be best to choose those pens that you can refill so that you can use the pen for a longer period of time.

Waste reduction is not only about the items going directly to garbage cans. One great means to lessen waste is to use resources wisely. Save water by taking shorter showers instead of taking longer baths. On top of that, this will also prevent the clogging of waste water in the sewage treatment systems. Also, compact fluorescent light bulbs emit light for a lengthier period of time compared to the usual incandescent bulbs that will be able to help in saving electricity and keeping the light bulbs away from landfills.

CHAPTER 3

DONATE OR SELL ITEMS INSTEAD
OF THROWING AWAY

Synopsis

When there are things that are no longer being used, most people end up throwing them away, without realizing that this action is another reason for the continuous increase of wastes in the world. In this chapter, you will realize that donating or selling items is a much better thing to do instead of throwing them away.

Anything that no longer serves you no longer earns a space in your life. However, one obstacle when it comes to de-cluttering that most people face is deciding on what they can do with the things that they want to get rid of.

Sad to say, you can easily get confused when facing this dilemma to the point that some people even get totally immobilized, and they end up failing to let go of their things simply because they can never decide where these should really go.

When it comes to reducing the number of wastes in the world today, do you know that you can help in your own way by donating or selling your stuff instead of merely throwing them to the trash?

Donate or Sell Your Things for a Cleaner Future

So, how do you know if you can still sell or donate your clutter?

When to Donate

All people have their own sense of identifying what can be donated. However, as far as donation is concerned, you can only give stuff to the different charities if:

- They still look appealing, nothing rusty, tatty and the like.
- It is still in proper working condition, with no need for any repair.
- It is still in decent quality, with nothing that is practically junk.

- It is in good condition.
- It can still be sold but you cannot bring yourself to sell it.

When to Sell

If you want to de-clutter, you need to consider some things before you can sell away your stuff.

- It is worth more than how much it would amount if you will organize a sale.
- You have the time to bother yourself on organizing the sale.

Donating and selling your stuff is not something that you should just do without mulling things over because if you just do it that way, you will end up making the condition of wastes in the world worse than when you started.

CHAPTER 4

REUSABLE BAGS

Synopsis

If you usually go shopping, do you know that there is something that you can do to help reduce the ever increasing number of wastes in the world? There are now reusable bags whose use is being promoted in many parts of the world for good reasons.

If you are just preparing to go to the grocery to do your shopping, what will you choose to bring - a plastic bag or a reusable bag?

When you use their plastic bags, do you know that it will take thousands of years for these bags to break down, which is virtually translated to forever? And during those years of slowly breaking down, these plastic bags will little by little separate to very small pieces of toxic particles which can contaminate both the soil and water that can then affect all the other living creatures, including yourself.

The Perks of Using Reusable Bags

Think about this. With all the problems that are associated to plastic and even paper bags, you will do yourself and the whole world a great favor if you will actually start to use the so-called reusable bags.

The benefits that you can get from using these bags are actually numerous and below are several reasons why you need to stop ditching the use of plastic and paper bags and invest instead in reusable bags.

Reduce Plastic Production

Petroleum and sometimes natural gas are required for producing plastic bags and surely, you know that these resources are non-renewable. Also, these can be very expensive and after some time, cost a lot of money. With the use of reusable plastic bags, you can lessen the production of plastic, thus reduce the amount of the non-renewable sources that will be consumed.

Empty the Landfills

The moment plastic bags fill up a landfill, these will take around 1,000 years before they degrade. The plastic bags do not biodegrade and instead, these photo-grade, meaning these are breaking down to toxic pieces. Thus, when you keep on using plastic bags, these will continue to pile up in the landfills, destroying the precious environment.

Protect the Wildlife and Marine Life

Do you know that plastic bags can affect both marine life and wildlife? In fact, more than hundreds of thousands of marine animals get killed yearly due to plastic bags. By using reusable bags, you can help preserve the lives of the other living creatures around you.

There are many other benefits that you can get out of using reusable bags. These are only some of them and there will be more that you can enjoy yourself once you start using these bags.

CHAPTER 5

LEARN WHAT IS RECYCLABLE IN YOUR AREA

Synopsis

Different areas have their own set of laws when it comes to the things that can be recycled. Make sure that you know the specific things that are recyclable in your life. In this chapter, you will get an overall idea of the materials which can be possibly recycled.

Most types of paper are recyclable. The prices of the virgin paper pulp are continuing to increase for the past few years and this is the reason why more and more plants are now being built in order to handle and reprocess waste paper.

Paper

What are the Things That You Can Recycle?

When it comes to recycling paper, it is important to collect large quantities of clean, well sorted, dry and uncontaminated paper.

Glass, Steel, Foil and Aluminum Cans

You can easily recognize these recyclables and in fact, many people never have any trouble when it comes to sorting the said materials properly. Steel, copper, aluminum cans, glass as well as other metals can be recognized right away and easily recyclable. But, you cannot mix glass bottles with other kinds of glass such as Pyrex plates, light bulbs, auto glass, glass tableware as well as window glass. As for aluminum cans, make sure that they are cleaned enough for preventing odors.

Plastic

There is a wide range of plastics that are recyclable and when the right dose of care is used, most of the plastic wastes will surely end up recycled. The main issue when it comes to recycling plastics is that you cannot mix various kinds of plastics during the process although virtually, it is hard to identify one type from the rest by simply looking at them or touching them. Even the tiniest amount of the wrong kind of plastic will already ruin the entire batch that will be reprocessed.

These are only the most basic of the things that can be recycled. While different areas have different lists of recyclables, you can expect that these will always be included.

CHAPTER 6

LEARN THE RULES IN YOUR AREA

Synopsis

Just like how there are different sets of recyclables in different areas, they also have varied rules when it comes to recycling. In this chapter, you will get to learn the most basic of all recycling rules that are being implemented in various places all over the world.

For many decades, recycling programs have already been in place that you will surely think that everyone is already aware of its pros. However, with all the labels, materials and rules, it is not a big surprise that many still end up confused. And with more and more communities adding different programs to their existing recycling services, another layer of confusion will surely arise.

Basic Rules in Recycling – Discover the Fundamentals

Good thing that good practices in recycling and composting always go together. Here is a quick glimpse of the simplest rules that will ensure that you will go on the right path of becoming a champion in zero waste management.

Plastic Bags are Not Allowed

If you will ask any recycle haulers, they will surely tell you that plastic bags are what they struggle with most. Even though these bags are recyclable, they go through an entirely different process compared to the hard plastics, wreaking havoc on single-stream/zero-sort systems which are common in most communities. As much as possible, you should try to s0rt out these plastic bags from your recyclables in order to help your community in its recycling goals.

Knowing Before Throwing

Education is always be the best way for you to learn the things that can and cannot be recycled and composted, saving the haulers from the hassle, cash and gas, not to mention that this will significantly help reduce environmental waste. It is important that you check with the hauler in your area to know the dos and don'ts when it comes to recycling.

No Throw When You Don't Know

This might seem a bit counterintuitive but the truth is, when you put items that cannot be recycled in the recycling bin, this will only lead to more harm to the environment than good. When in doubt, it is always a better idea that you avoid putting the material with the rest of the recyclables.

CHAPTER 7

SEPARATE YOUR TRASH

Synopsis

Putting all your trash together can result to some serious damages to the environment. Trash separation actually plays a crucial role in ensuring that no further harm will be brought to the environment than what it is suffering from right now.

If there is one important environmental knowledge that you should know, it is the fact that trash separation actually helps lessen the environmental pollution, turning your wastes into priceless treasures.

Trash Separation – Wonderful Benefits That You Never Expect

The truth is, waste separation helps when it comes to reduction of the waste volume, processing equipment, consumption of the land resources, lower costs on processing together with some ecological, economic and social benefits.

The very first benefit of trash separation is the fact that this can lessen the number of areas that will be consumed for landfills. During recycling that is made easier by separating those that can be recycled from those that cannot be, the volume of wastes can be lessened to as much as 50%, thus decreasing the need for more landfills.

When you separate your trash, you are also helping reducing the environmental pollution. Batteries put into waste contain cadmium, metal mercury as well as other substances that can be toxic and cause some serious harm to the humans. Meanwhile, plastics can cause failures of crop planting, with animals ending up eating plastic waste that can result to their death. But through proper separation of trash, these hazards can be reduced significantly.

Separating your garbage can also pave the way for you to enjoy getting cash out of your trash. In fact, there are more and more countries that are now developing their own systems when it comes to garbage collection and ensuring that useful items will be made out of those that can still be recycled.

CHAPTER 8

BUY RECYCLED ITEMS

Synopsis

If you are not in favor of the idea of recycling items, you can still help in the cause by purchasing the recycled items instead. Discover in this chapter how patronizing items made from recyclables can make a big change in both your life and the life of the entire planet.

Among the most immediate challenges in recycling today is getting the consumers educated regarding the benefits that they can enjoy once they purchased the products created using recycled materials. The marketplace drives manufacturing and purchasing products that contain recycled materials establishes a long term market for the recyclable materials, increasing the revenues of recycling programs. Once the demand continues for the recycled products, the manufacturers will then produce more recycled products.

Patronize Recycled Items for Everyone's Benefit

The great benefits that you can enjoy when your purchase recycled products are as follows:

➢ **Save energy** – Usually, it takes lesser energy for creating recycled products. For instance, the production of aluminum coming from recycling takes 95 percent less energy compared to producing brand new aluminum coming from the bauxite ore.

➢ **Save the natural resources** – When you choose to buy recycled products and not those that contain virgin materials, you are actually considering the land, reducing the need for you to dig for more minerals, harvest trees and drill for more oil.

➢ **Save clean water and air** – Instead of using virgin materials, the use of recycled materials will lessen the chances of getting pollutants eliminated during the acquisition, processing and manufacturing of the product.

➢ **Save landfill space** – Once the materials that have been recycled went to new products, these will no longer need to be dumped to landfills, thus saving space.

➢ **Create jobs and save money** – The process of recycling can create more jobs compared to incinerators or landfills and usually, recycling is the least expensive method for waste management being used by different towns and cities.

CHAPTER 9

WHAT TO AVOID

Synopsis

Similar with most aspects in a person's life, there are still some things that you need to stay away from to make sure that your recycling goal will be achieved in the best way possible.

The Big No No's In Recycling

Recycling might seem to be a rather straightforward system but similar with other processes, there are still some shades of gray in every black and white as far as the designations of paper, plastic and glass are concerned. There is one important rule that you must always remember and that is the fact that not all plastic, glass, and paper can actually be recycled in an equal manner.

Here is a quick rundown of the non-recyclable items that you need to avoid:

- ➢ Brightly dyed paper since the colored ink can actually leak and change the other items' color.
- ➢ Paper towels and napkins are deemed unsuitable in the process of recycling since these can absorb some contaminants.
- ➢ Cardboard containers coated with wax and juice boxes – Once these do not have the mark of being recyclable, you should not recycle them at all.
- ➢ Plastic screw-on tops – Even though you can recycle plastic bottles, the tops are not regarded as appropriate for recycling.
- ➢ Wet paper is also not recyclable due to the damage to fibers and possible contaminants.

CHAPTER 10

THE BENEFITS OF RECYCLING

Synopsis

Now that the book has finally come to an end, this is the perfect time for you to finally get to learn the benefits of recycling, not only for yourself but more importantly, for the people around, you the society you belong to and the Earth that you are living in.

Recycling Benefits – The Effort is Worth It

Recycling is definitely the most common symbol that you can see in trash cans, garbage bags as well as dump trucks. Children in various parts of the world are taught of the phrase "Reduce, Reuse, Recycle" hoping that a cleaner environment will be created in the near future. Everyone knows that even in the simplest way, all people can actually help create a much better place to live in. However, there is no denying that there are plenty of benefits that you can get out of recycling aside from lessening the amount of trash that you need to get rid of off every day. After all, both effort and time are used up for the collection, separation as well as sending away of trash.

Conserve and Preserve the Natural Resources

Old bottles, scrap cars, used rubber tyres, and junk mail have now become common adornments in landfills. These are probably endless in number but the required resources will finish off rather quickly. Through recycling, everyone gets the chance of letting the junk items be used time and time again in order to avoid exploiting the new resources. This will also help conserve the natural resources including water, coal, oil, gas, minerals and timber. One more benefit of recycling is that this will put more emphasis on the creation of technology. This is the reason why majority of industries are supporting the programs wherein they can gain large amounts of recyclable materials that can be converted to new items.

Save Energy

When recycling aluminum cans, you will be able to save as much as 95% of the entire energy needed for producing the cans from the raw materials.

You can say that the energy that you can save from recycling is enough for you to shed some light on a bulb and make this last for 4 hours. It has been clearly shown that there is a high amount of energy that you can save and use in the future, thus reducing the reliance on foreign land that can futther help in saving money along the way.

Reduce Landfill Size

Among the primary reasons why recycling is being promoted is the fact that this can reduce the environment's strain. Through the constructive use of waste products, you can little by little lessen the size of landfills.

More Opportunities for Employment

If you think that recycling is only for his or herself, you better think twice because the truth is, it is already a big industry even all by herself. After doing the basic sorting out and depositing trash for recycling, it is important that these are sorted out and shipped primarily off to the correct places.

Offer Some Cash Benefits

It is a must for you to know that recycling is not something that is merely for the purpose of being charitable much less doing the good things that will help you preserve your environment. If this is really the case, everyone is definitely going to recycle out the goodness of their own hearts. Majority of governments have some policies put into place that will give some financial benefits to the ones who do not really recycle. Those people who are taking glass bottles or aluminum cans to the plant that recycles usually get some cash in return. The truth is, a lot of teenagers now pick up recyclable items for them to make some extra money during their free time. Appliances, old newspapers, steel, copper, plastic as well as beer cans can also be the source of money.

Reduce the Emissions of Greenhouse Gases

When recycling products, there is always the tendency of saving energy that can cause less emission of the greenhouse gases. These gases are the ones responsible for the increase in the case of global warming. This can help lessen water and air pollution through cutting down the primary source of the pollutants which are being release to the environment.

Save More Money

If there is an expected place where you can see recyling's benefits, it will be none other than the economy that is strong and efficient. One of the many things that can drag down the good things about this is the use of some sensitive products. At the end of the day, it is important to know that once the jobs increase, there will also be a boost in economy. Once the cost of sustaining the present waste disposal go down, every single money that you have worked for will be given to those who really need this the most.

Stimulate Usage of Greener Technologies

Through recycling, this will further motivate people to go for greener technologies. The use of the renewable sources of energy such as wind, geothermal or man is now on the rise that can help a lot in conserving energy and reducing pollution.

Bringing Various Communities and Groups Together

When the day comes to an end, recycling is one process that brings together even the entire community. Even if it is as simple as picking up a candy wrapper on the road or collecting the waste materials for raising money for colleges and schools, all of these are traits shared by each and every member of the family. Even the simplest programs can make communities much stronger than ever, especially when they are built on the numerous benefits of recycling.

While it may seem at first that the benefits of recycling are simple, little by little, you will notice that great changes can be made out of the simple determination of creating a better world for your children and their children.

Printed by Libri Plureos GmbH in Hamburg, Germany